The Earth Is Not Flat

by
Gordon S. Brooks

Copyright © 2016
by Gordon S. Brooks
All Rights Reserved

To Karen, who lets me go off on the most bizarre tangents and still loves me

Contents

Introduction 1

Flat Earth! Wait, What? 5

If Not a Globe, Then What? 9

Proving the Flat-Earth 13

I Object! 19

The Grand Conspiracy 27

Flat-Earth Tactics 31

Flat-Earth Baggage 39

Addressing Flat-Earth "Proof" 43

The Lunatic Fringe 57

Perspective For Flat-Earthers 79

Flat-Earth Morals 85

Why It Matters 91

Introduction

You might be forgiven for assuming that the title of this book is a metaphor for the poor state of scientific literacy in the world today, because, for all that we, as a society, don't know about science, even as we enjoy its fruits on a daily basis, no one really believes that the Earth is flat.

Alas, no such luck. No, really; it seems there is a rather large and growing number of people who actually believe that the Earth is flat, and they are using the Internet, YouTube in particular, to spread the worst kind of misinformation to pull in gullible people and convince them that, despite everything they've been told, the Earth is not a globe.

I first encountered these people sometime in the middle of 2015, on YouTube, when I was searching for something I don't exactly recall. I was flabbergasted. Not that there were people who believed such things, but that there seemed to be so many, including some who had written books, and held incredibly long Google hangouts, and spent a great deal of time hashing out what they claimed to be the finer points of their arguments in the comments.

Now, this is that point in the book where you expect me to say, "but the closer I looked, the more I was convinced that they were on to something, that there was real evidence that the very nature of the world we live on is not what we've been taught since childhood."

But I'm not going to say that. Because the fact is, the only way to buy into this nonsense is to not think about it too much or look into it too deeply. And that shallow understanding is precisely what the flat-Earthers on the Internet demonstrate time after time.

But why should it matter? Why do I care? People can believe anything they want, after all, and certainly the flat-Earthers represent a very small number of people in the world. Really, what's the big deal?

To a certain point, I agree. But it bothers me because I see it as a growing trend toward anti-scientific thinking. This at a time when we rely on science and technology more than ever, and yet we seem to understand less and less of it.

As Carl Sagan said:

> We've arranged a global civilization in which most crucial elements profoundly depend on science and technology. We have also arranged things so that almost no one understands science and technology. This is a prescription for disaster. We might get away with it for a while, but sooner or later this combustible mixture of ignorance and power is going to blow up in our faces.

To me, the flat-Earth "movement" is a very visible indicator of just this kind of problem. Because it's not just that some people believe in a flat Earth; it's a combination of all the other things they believe, the things they don't believe, and the things they don't know that I find worrying.

Even more worrying is that, with easy access to scientific information without the requisite scientific understanding, these people are able to give their beliefs the trappings of science, even to the point of claiming that they are doing "true science" while working scientists are just following dogma. And so they sound more intellectual and convincing than their facts and arguments warrant.

This applies, of course, not only to flat-Earthers. We've

seen this with the creationists time and time again, and it seems that the creationist movement is much more dangerous, as it has a lot more political power. So why focus on the flat-Earthers? Because they cover a wide range of odd and interdependent beliefs that I have come to see as a microcosm that reflects the wider problem of pseudoscientific belief. And while pseudoscience has been with us at least as long as science, in the age of the Internet spreading falsehoods is easier and more effective than ever before in human history. For those not well-versed in science and skeptical of authority, or predisposed to certain fundamentalist religious beliefs, it's a potentially dangerous trap.

Flat Earth! Wait, What?

Hold on, you say, didn't Columbus settle the whole flat Earth thing in 1492? Well, actually, the whole story about people in the 15th Century thinking that the Earth was flat is a myth, made up by, among others, the American writer Washington Irving. The idea that the Earth is a sphere is about 2600 years old, perhaps older, and was pretty much a settled question in most of the world well before Columbus set sail for what he thought was India. So, yeah, we've been done with the whole flat Earth idea as a civilization for a very long time.

But there has probably been a fringe belief in a flat Earth for almost as long as the globe Earth idea has been around.

The current crop of flat-Earthers can trace their philosophical roots, if they deserve to be called such, to an itinerant lecturer known as "Parallax," whose real name was Samuel Rowbotham. Among other less-than-scientific pursuits, Rowbotham made his living talking up the idea of a flat Earth. He also wrote a book about it called "Zetetic Astronomy: Earth Not a Globe." (The actual subtitle is much, much longer, but from here on out, I'll refer to it simply as "Zetetic Astronomy" to save myself an inordinate amount of typing.)

Today's flat-Earthers draw most of their basic arguments from Rowbotham, and sing the praises of his experiment at a waterway called the Old Bedford Level, in which Rowbotham basically had someone get into the water with a telescope and look at a boat six miles away, and since he could see the entire boat, Rowbotham concluded that the Earth did not curve. His math was a little off, but the real problem with the experiment was that Rowbotham's telescope was only eight inches above the water, and Rowbotham did not account for refraction *even though he mentions it earlier in the book.* My conclusion, for which I'm sure I will catch fire from the flat-Earthers, is that Rowbotham rigged the experiment, and knew full well that he was selling his audiences a bill of goods.

But I have no more proof for that conclusion than the flat-Earthers have for their conclusion that the Earth is flat. Oh, yes, they have proof, or at least something they call proof. I'll get to that shortly, but it's easier to evaluate this so-called proof if you have some idea of what the flat-Earthers claim is the true nature of the world we live on.

If Not a Globe, Then What?

I think that when most people picture a flat world, they think of comedic illustrations of a square, with water (and often ships) falling off of the edge. But for the most part, modern flat-Earthers aren't proposing anything like that. The most common model of the flat Earth looks like the image on the next page.

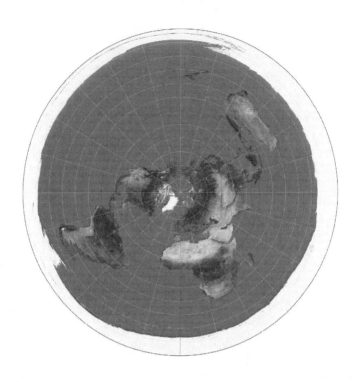

People who are very familiar with geography or cartography should recognize this immediately: it's an azimuthal equidistant projection of the Earth, with the North Pole as the origin point (which I'll refer to as the AEP). If you're not familiar with geography, let me tell you what you're looking at. A projection is a way of translating a globe Earth to a flat map. The most familiar projection is the Mercator projection, which is hanging in a lot of classrooms, and has rightly come under fire lately for distorting the areas of countries that are not along the equator. But, there's the rub: any projection is going to involve some kind of distortion; it's just a consequence of representing a three-dimensional object in a two-dimensional medium. And the AEP is no exception; the east-west size of Australia, for example, is more than doubled.

Yet, many if not most flat-Earthers propose this as an accurate representation of the Earth. How do they explain the distortion? I'll get to that, and you probably won't believe it when I do.

And what of the edge? Well, flat-Earthers say that there isn't one, not in the way we think of it anyway. Antarctica, you see, is not a land mass at the bottom of a sphere, but a great circle, an ice wall that surrounds the outer ring of the Earth, just as depicted in this projection. What's beyond the ice wall? Well, opinions vary. Many, maybe most, say that it ends at the bottom of a vast dome, which is where the stars and planets are. Others say the ice goes on infinitely in all directions, although where that leaves us with the stars is a little vague in that case.

The sun, they say, is not 93 million miles away, but a mere few thousand, and only 36 miles across. The moon is at the same distance and size, which conclusion they reach by looking at the apparent size of both. The two lighted objects are traveling in a circle over the equator (though not always; the sun moves between the Tropic of Capricorn and the Tropic of Cancer to make the seasons happen), giving us alternating day and night.

That's the gist of it. There are variations, of course, even arguments, but this is what the majority of flat-Earthers I've run across think the world that you and I live on looks like. It's nearly the model proposed in *Zetetic Astronomy* (though Rowbotham was more careful and corrected the land mass distortions, opting for distorted distances between the continents instead). It's what flat-Earth proponent and hardcore Christian fundamentalist Wilbur Glenn Voliva firmly believed. Despite the fact that the origin of this map is as a projection of what its inventor, at least as long ago as the 11th century, undoubtedly knew was a globe, this is what most flat-Earthers are defending when they try to prove that the Earth is flat.

Your mind is probably already dancing with questions, about how this is all supposed to work, and what about people who've circumnavigated the globe, and satellites, and those pictures we have of the Earth from space. I'll get there, I promise. But first I'm going to take you on a little journey through fantasy land, to the flat-Earth version of what constitutes proof that we are not living on a globe. Hang on, it's going to be a bumpy, confusing, and occasionally infuriating ride.

Proving the Flat Earth

What if I were a proponent of the flat Earth, and I was trying to convince you, who had known all your life that you are living on a globe, that it had all been a lie, and that the world was flat after all? Well, going with the strategy used by current flat-Earthers, I would try to do it by disproving the prevailing model, the globe Earth. In 1885, after the death of Rowbotham, William Carpenter published "One Hundred Proofs That the Earth Is Not a Globe." Not to be outdone, in 2015, flat-Earth proponent Eric Dubay published "200 Proofs Earth Is Not a Spinning Ball." Note that neither title says "Proof That the Earth Is Flat." A subtle difference, but in the long run it makes a big difference.

But let's take this approach, and try to convince you that the Earth is, in fact, flat. Every argument I'm about to use has been used at least once (and likely many times) by a promoter of the flat Earth.

To start, I would tell you that you only believe the Earth is a globe because you have been told, because your teachers told you, and the scientists told you, and the government told you, that you live on a big ball that spins at over 1000 miles per hour, and hurtles through space at over 67,000 miles per hour around the sun. But, I would say, you have no direct evidence of this. And all you have to do to open your eyes is to look around you.

I would say that no matter how high you go, whether at sea level or Mount Everest, the horizon always looks perfectly flat, 360 degrees around you. And that all footage from planes and weather balloons, unless distorted by fish-eye lenses, shows a flat horizon. In addition, the horizon always rises to eye level, no matter how high up you go. If the Earth was a globe, even a big one, you would see the horizon drop below your eye level, and you would have to look down to see it.

And then there is the matter of water. Water, we know, always seeks its level. A calm, flat lake, or even the ocean, could never curve around the surface of a globe. If it did, then rivers like the Nile and the Mississippi would be flowing uphill.

Engineers who build railways and roads never take the curvature of the Earth into account. If I were trying to convince you, I could find quotes from engineers who find complete folly in any notion of allowing for curvature. Even a quote from the Manchester Ship Canal Company:

> **It is customary in Railway and Canal constructions for all levels to be referred to a datum which is nominally horizontal and is so shown on all sections. It is not the practice in laying out Public Works to make allowances for the curvature of the earth.**

I would make much of the distance one can see over bodies of water. If the Earth curves away at 8 inches per mile squared, we should not be able to see lighthouses from hundreds of miles away. We should not be able to see the Chicago skyline from across Lake Michigan, 90 miles away. Heliographs, used in the late 19th Century as signaling devices over long distances, should not work on a curved Earth, as the horizon would block the line of sight.

If the Earth were round, I would say, the airplanes would have to constantly fly with their noses down to keep from flying off in space. In fact, the gyroscopic systems used in airplanes would be useless on a globe, since gyroscopes maintain their orientation in space without regard to gravity or curvature.

If the Earth was spinning at a thousand miles an hour, I would propose, then cannonballs fired straight up would fall back to Earth miles away from the cannon, instead of nearly perfectly back into the barrel, as happens in reality. Helicopters and balloons could hover in place and travel thousands of miles without any propulsion whatsoever. Airplanes would crash attempting to land on swiftly-moving runways. In fact, they would never reach their destinations at all in any eastbound flight, and they would travel three times as fast going westbound.

In a recent space jump, Felix Baumgartner could not have come close to his target had he been jumping toward a rapidly-spinning ball.

In fact, I would emphasize, if the Earth were spinning at 1000 miles per hour, you could not stand on your feet. Surely, the argument goes, you would feel it if you were moving that fast.

If the Earth curved away from you, then buildings you see in the distance would visibly lean away from you, and yet no one sees this.

I would point out several anomalies in the sky. Like the fact that during some lunar eclipses, the eclipse can be seen starting while the sun is still in the sky. And that when the moon and the sun are in the sky at the same time, the bright side of the moon often does not appear to be in a proper line with the sun. I would state that several constellations can be seen from both the Northern and the Southern hemispheres, which is clearly impossible if the stars are hidden behind the enormous bulging curve of the spherical Earth.

I would also take note of the fact that when you see the sun or the moon on the horizon over water, the reflection forms a line from the light source toward you, which should be impossible on a curved surface.

I might point out, too, that if the Earth is rotating every 24 hours, and orbiting the sun every year, then every six months, day and night should be reversed. Think about it.

And I would be remiss if I did not include the experiment by Rowbotham on the Old Bedford Level, in which Rowbotham clearly saw a boat six miles away which should have been hidden under 27 feet of curvature.

Clearly, under the weight of all this evidence, you would at least have to concede the possibility that the Earth as you know it might all be a big lie. Right?

I Object!

Foul! you say. Even if all of this is true (and we will get around to that question down the line), it doesn't prove that the Earth is a flat disc with the North Pole at its center and a great ring of ice as its perimeter. And it doesn't negate any of the evidence we have that the Earth is, as we have been told, a big spinning ball.

Good point. Let's see how the flat-Earthers explain some common objections to their Earth model.

Let's take something as simple as day and night. How, on the flat Earth model, can there be day on part of the Earth and night on the rest? The flat Earthers will point out the the sun, circling over the equator, is very close, only a few thousand miles in altitude, and thus only casts light on part of the planet. So, within the reach of the sun's light is day, and beyond the range of the sun's light is night.

They have even made computer animations of this, though they do not show half the disc in light and half in darkness. Well, at least most of them don't, but some of them do, admittedly without any explanation of how this is possible. The nature of the sun's light, they say, is unknowable.

Even if you buy that the sun can somehow light exactly half the disc, you may wonder how the sun can disappear from view for half the time. After all, if it's thousands of miles up and only thousands of miles distant, it should still be visible, shouldn't it? The flat-Earthers will variously talk about light only reaching so far, and the limits of human vision. But specifics (and experiments) are hard to come by.

Since we're talking about the sun, what about sunsets? If the sun is circling overhead, how would we get a sun that seems to sink below the horizon in a westerly direction when it sets, and then show up in an easterly direction when it rises the next morning? Proponents of the flat Earth have one go-to explanation for this illusion: perspective.

I'll have a lot more to say about flat-Earthers and perspective later, but here's the short version for sunsets: perspective makes the ground rise up to eye level, and the sky come down to the horizon line, and that's what makes the sun seem to set. They will show video of the sun getting smaller as it sets, and offer that as proof that perspective is at play. If you don't agree that this is how perspective works, they will scoff at you and tell you you do not understand the "laws of perspective." However, when asked for a source for the "laws of perspective," they will produce either other flat-Earth sources, or complete silence.

And they won't even talk about the direction of the sunset. If you're facing west, from anywhere on the Earth, in the flat model the sun will veer severely off to the north before it "sets" (which, in fact, it would never do). This is, of course, nothing like what we see in reality.

And since we're on the subject of the sun, let's have a little talk about the moon. Of course, there is the same problem (and the same answer) to the question of the moon rising and setting. But to that we need to add questions about the phases of the moon, and eclipses. For phases, flat-Earthers will construct elaborate computer models that seem plausible rendered from above, but somehow they are never rendered as if seen from the ground. And eclipses are supposed to involve some dark object that blocks the light of the moon, though the nature of this object and how it works to make the eclipse appear as it does, are quite vague.

Most flat-Earthers also believe that the moon shines by its own light, rather than light reflected from the sun, though this does no more to explain moon phases and eclipses than any other hypothesis.

Solar eclipses are hardly mentioned, though it seems that the same kind of mechanism must be involved.

Okay, then, let's get back to Columbus, who sailed west instead of east to try to get to India, and whose biggest mistake in that venture was not taking Eratosthenes at his word about the circumference of the Earth. Magellan, of course, is the first known to have circumnavigated the globe. What do the flat-Earthers say about that?

Well, they say that these voyages were taken around the circle of the disc, not around the globe. The distances don't quite work out, at least not on their version of the map, but they have a point in that you could make such a voyage without reaching the blockade of the great circle of ice that is, they say, Antarctica.

It would be different, of course, if somebody had actually circumnavigated the globe from pole-to-pole, and crossed Antarctica. The fact that someone has is something that the flat-Earthers simply refuse to accept.

Just as they refuse to accept the existence of satellites. GPS and dish TV, they say, work with ground-based systems, using cell towers and other big antennae, or by skipping signals off the ionosphere or, for some, the "dome." They will point to dishes that are pointed almost level to the ground as proof, ignoring the fact that, the closer you get to the equator, the higher the dishes point, with dishes pointing pretty much straight up in equatorial regions.

To support the absence of GPS satellites, some flat-Earthers will point to flight tracking sites, and note that planes flying over the southern oceans will disappear from tracking unless they are within a couple of hundred miles of a ground station. No matter how many times it is explained to them that the satellites are not tracking the flights (GPS is a receiver-based system; the plane receives GPS signals and then transmits its position to ground stations), they will still hold this up as an example of proof positive that there are no satellites.

What about the distorted sizes of Australia and the southern regions of South America and Africa? There's a lot of hemming and hawing about this. Some say that no one really knows the true size of those continents, which must be a surprise to anyone who's driven or flown across them. There are some who say that planes flying from one of the southern continents to the other always go through the Northern Hemisphere, which is simply not true. Others propose that jets are actually capable of flying much faster than we've been told, and that's why it's possible to fly between those southern continents in what seems like a normal amount of time.

In short, the explanation for the wider-than-expected east-to-west length of the regions far to the south is no explanation at all.

Which brings us to to the most damning piece of evidence against the flat Earth, and in favor of the globe we all know and love. We have pictures.

On December 22, 1968, the crew of Apollo 8 took the first picture of the entire Earth from space. This was not the famous "Blue Marble" photograph taken by the last mission to the moon, Apollo 17, a lucky shot of the Earth with the sun at the crew's back, showing one entire lit side of the Earth. Earth's first portrait had about three quarters of its face lit, with the sun to the astronauts' left.

Apollo astronauts took a lot of pictures of their home planet during the ten missions that left low-Earth orbit and went to the moon. Surely, it is obvious from these photos that the Earth is, as we've known for millenia, a sphere, lit by and in orbit around the sun, with its single natural satellite, the moon. What do the flat-Earthers say to this?

Simple, for them. It's all fake. The photographs from the 60s and 70s are actually paintings, and more recent pictures from satellites are actually nothing but computer-generated imagery. Humans never went to the moon; it was all a show, put on by NASA (and directed by Stanley Kubrick, apparently) contrary to everything we know about film and video technology at the time. In fact, man has never even ventured into space for, you see, in the flat Earth model, what we know as outer space doesn't exist at all.

I hope you can see a pattern emerging here. In order to support the flat Earth, no matter if you believe that the AEP is accurate, or if you think there is a dome, or if you agree with the sun circling around the equator, you have to believe one thing for certain: that a very large number of people are lying to you.

The Grand Conspiracy

To flat-Earthers, the globe isn't just a mistaken notion by scientists. It's the big lie. It's the conspiracy of governments and scientists all over the world to keep us from learning the truth. What is this truth? Other than that the world is flat, I mean.

That depends on who you talk to, but there are three approaches that I hear most often. One is that if the government, through its schools, can lie to you about something as basic as the shape of the world you live on, then it can and does lie about everything, as a means of control. Another reason for the big lie is that vast natural resources can be found beyond the Antarctic ice wall, and the governments and big corporations don't want us to know about it so that they can reap all the profits for themselves and keep us in poverty and, again, under control.

The third reason speaks to the root of the flat Earth belief for many. This group believes that the Earth is flat because they believe that the Bible (or the Qur'an) says so, and no amount of physical evidence will shake that belief. If the evidence says that the Earth is not flat, than the evidence, or our interpretation of the evidence, must be wrong. And the world's governments are lying about the shape of the Earth to separate us from God.

But this isn't just a government conspiracy. In order for this to work, the airlines have to be in on it, all of the scientists, sailors, and all those people who build, maintain, and install satellite systems. Not to mention Google and every cartographer in the world. The conspiracy is variously (and sometimes simultaneously) ascribed to the Masons, the Illuminati, and the Zionists (or just Jews in general). Flat-Earthers tend to also be believers in chemtrails and 9-11 conspiracy theories. In an odd twist, some flat-Earthers believe that the recent rise in popularity of flat-Earth topics is actually a CIA psychological operation (psyop) to discredit the flat-Earth movement. (There are, of course, people who don't believe in the flat Earth who also believe that the flat-Earth movement is a CIA psyop to discredit other "truthers.")

This kind of paranoia causes a lot of rifts in what is sometimes called the flat-Earth community, and anyone who ditches the flat-Earth theory over any trivial thing like contrary facts is instantly accused of being a government shill. As is anyone who comments on flat-Earth videos or forums and points out the flaws in the theory. This is one of the tactics used by flat-Earthers. There are, as you will see, many others.

Flat-Earth Tactics

Usually the very first response you get when you challenge the flat-Earthers in the comments sections of YouTube or other social media, if you are encountering the flat-Earth concept for the first time, or if the flat-Earthers involved are encountering you for the first time, is that you just haven't done your research. You are the product of a lifetime of indoctrination, and if you just looked into the idea with an open mind, you would soon find yourself convinced that the Earth cannot be a globe.

There are multiple conceits at play here. One is that, by accepting a couple of thousand years of solid scientific inquiry as very likely factual, that you are being closed-minded. And since no one who values the quest for knowledge wants to be thought of as closed-minded, this can be very effective at roping people into looking at the evidence.

Which leads to the next conceit, that there exists a vast body of untapped evidence leading to the truth of the flat Earth. This is hinted at by another popular flat-Earth tactic which I call the numbers game. The first volley of this play is referring a newcomer to DuBay's 200 proofs, which sounds like an overwhelming number until you actually read them and find that most are repetitive, and all are easily dismissed with just a small amount of critical thought.

The next volley of the numbers game comes when, in a comments section somewhere, you back a flat-Earther into a corner on some point of fact, and they respond with the litany of standard comebacks, which usually include the Michelson-Morely experiment, Airy's Failure, and the Sagnac Experiment. I'll have more to say on those (and other items in the litany) in a bit, but this spouting of the standard list is part of another conceit: that the flat-Earther you are arguing with has done deep research into the subject, and is thus an expert, someone to be reckoned with.

But the very act of regurgitating this list demonstrates a profound lack of research. Citing those experiments as proof that the Earth is stationary shows a failure to read the actual papers involved instead of another flat-Earther's explanation of them.

It doesn't take any deep research to discover that Airy's Failure was a failure to find the aether, the substance widely believed to inhabit space and the medium through which light was thought to travel, until late in the 19th Century. Michelson-Morely was another nail in the aether coffin, and paved the way for the theory of relativity. And Sagnac's experiment lead to the discovery of the Sagnac Effect, which is used to make navigation systems that compensate for the Earth's rotation, exactly the opposite of what flat-Earthers claim.

You don't have to take my word for it. A few minutes of well-phrased Google searching will give you enough information to shoot down the flat-Earth litany. I'll return to the litany in a bit, but I was talking about tactics.

Another popular tactic of flat-Earthers and other pseudoscientists is selection. This includes selecting only information that supports their ideas while ignoring information that doesn't, and also reporting only a portion of the information that they claim proves their case.

A glaring and common example of this is footage on YouTube showing distant skylines that, purportedly, we shouldn't be able to see over the Earth's curve. Sometimes distances are given, sometimes not. The height of buildings in the skyline are rarely given, and most often when they are, it's the height above ground level, not above the water level.

The height of the observer is nearly always estimated, and badly, judging by the foreground visible in the footage. These things matter a good deal.

Most damning, refraction is dismissed out of hand as a reason for possibly seeing just a bit further than the math would seem to indicate. Even Rowbotham acknowledged the effects of refraction, although he conveniently ignored them at the Old Bedford Level.

As an aside, the flat Earth math for how much of something you can see over the curve is also faulty; their eight inches per mile squared is an old surveyors' rule of thumb for the drop in height due to curvature. It assumes the transit and target are the same elevation. The math has to involve the height of the observer and the target, or it's meaningless.

Which brings me to another version of selection: invoking a proof when it supports your position, and ignoring it when it does not. Math, for example. They are adamant that if you can see more of a distant skyline than the math indicates, the Earth must be flat, and ignore the obvious fact that any amount of occlusion indicates that the curve is present. But if you invoke math to show that a sun circling over a flat disc would never appear to set, then you're just using abstraction, for math doesn't describe the real world.

Handy, huh?

Refraction is another of those factors that come and go in the flat Earth. It can't be used for to explain the viewing of distant skylines, but it can be used to explain sunsets. You get the idea.

Another flat-Earth tactic, which goes at least as far back as Rowbotham, is out-of-scale illustrations. Now, illustrations that are exaggerated or simplified have their places, as long as the aspect of the illustration you're exaggerating doesn't materially affect your argument. Consider this illustration from *Zetetic Astronomy* depicting the Bedford Level experiment.

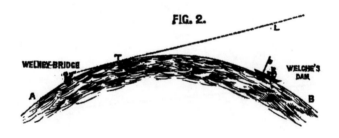

It is painfully obvious that this illustration is so blatantly distorted that it can have little meaning. And yet not only is this same tactic used repeatedly in flat-Earth presentations, but this illustration, and others like it from Rowbotham, are reproduced as evidence to this day.

When faced with a question they can't satisfactorily answer, many flat-Earthers will change the subject. I can't count how many times I've been commenting on perspective, only to find myself asked to defend the moon landing, or explain why all the pictures of the Earth since 1972 have been composite images, something I'm not planning to go into in detail in this book (if you're interested, you can read up on the history of space travel for yourself).

The most daunting method of changing the subject is presenting a barrage of questions or proofs, usually copied and pasted, into the comments, implying that if you don't take the time and energy to answer every one of these, you are conceding defeat and admitting, by default, that the Earth is, in fact, flat, stationary, and the center of the Universe.

In defense, we who comment on flat-Earth videos and forums, have our own copy-and-paste responses just to save on finger cramps from typing the same things over and over.

Among the last tactics tried by flat-Earthers is outright fabrication. They will invoke heretofore unknown laws of nature concerning the behavior of light, electromagnetic fields, and, especially, the non-existence of gravity.

Yes, gravity. It's actually very hard to sustain a flat Earth in the face of gravity. More on that momentarily.

When these tactics fail to bring about the conversion experience, the flat-Earthers will go on the attack. They will attack your religious conviction, if that is their basis for believing the flat Earth. Young-Earth creationist Kent Hovind has been plagued by flat-Earthers for his refusal to accept the "Biblical truth" of the Earth's lack of sphericity.

They will call you a troll, or a shill, in many cases having no idea what these supposed insults actually imply. They will attack your character, your age, and even your personal appearance. Then they will delete your comments and block them. Nice exchange of ideas, there.

There have even been threats. Not me, really, but I have personally seen them directed at others.

Why go that far to promote an idea for which you have no good evidence? Well, zealotry, of course. We've seen that for thousands of years. But also a good dose of paranoia. You'll see more of that in play near the end of the book, when I talk about the true fringes of the flat Earth.

Of course, believing in a flat Earth brings some necessary baggage with it. There are lots of ways that people picture or explain the flat Earth, but there are some things you really have to believe if you want to keep believing that the Earth is flat.

Flat Earth Baggage

First, as I said, in order to believe that the Earth is flat, you have to buy into the vast conspiracy; an awful lot of people in the world have to be lying about pretty much everything they spend their days doing. But you also have to believe that spaceflight is impossible, unless somehow it's possible but it's all faked. You have to believe in some warped version of the propagation of light to make the moon and the sun work. You have to believe in some unwritten law of perspective to make sunset and sunrise work.

You have to believe that gravity is a lie. Because if the law and the theory of gravity are correct, a flat Earth is completely impossible. Anything as massively big as the

Earth (and even the flat-Earthers can't deny that it is really, really big) would pull itself into a sphere by way of gravitational attraction. There's just no way around that without discarding gravity.

And, indeed, many flat-Earthers do just that. Some say that density and buoyancy can substitute for gravity, which shows a lack of understanding of physics that boggles my mind when I consider that these people are claiming to be more knowledgeable than all the physicists in the world. I have suggested, unkindly, I admit (in a fit of frustration), that these flat-Earthers try a little experiment: take a one-inch steel ball and a three-inch steel ball, which have equal density but a different mass, and drop each in turn on your foot. Which one does the most damage? Which hurts more? Gravity at work!

Aside from this simple and satisfying experiment, there are many ways to prove gravitational attraction, experiments done by physics students the world over. There are even sensitive instruments called gravimeters, used to measure tiny variations in gravitational fields, which seem to work quite well for the geophysicists and miners that use them, unless they, too, are in on the conspiracy.

A very clever fellow on YouTube posted a video that pretty much makes the case for gravity right from your fingertips: all modern smartphones and tablets have sensors that detect gravity. Otherwise, there is no way the device would be able to adapt the display to the proper orientation when you rotate the screen. I admit I hadn't thought of that. Brilliant!

DuBay even mentions gravity as early as "proof" #23 in his little book, where he puts the word in scare quotes. In his world, gravity doesn't exist, and can't hold the oceans or the atmosphere. In proof #116, he states:

> **There has also never been a single experiment in history showing an object massive enough to, by virtue of its mass alone, cause another smaller mass to orbit around it.**

I suppose I should reiterate at this point that another thing you can't believe in if you're a flat-Earther is satellites.

There are a few flat-Earthers who claim that Earth's gravity is created by acceleration. In other words, this flat disc we supposedly live on is accelerating upwards at a rate of 9.81 meters per second per second, thus creating the very gravitational pull that we all know and love.

They fail, though, when asked two questions (which I've yet to see them answer). First, if the gravity is due to constant acceleration, why is there less pull from gravity at higher elevations? Second, how fast are we going? That second question is sometimes answered, "I told you, 9.81 meters per second per second." But that's acceleration, not speed. If the Earth/disc was constantly accelerating, even if the Earth was 6000 years old (yes, a lot of flat-Earthers are also young-Earth creationists), then the Earth would currently be traveling at somewhere in the neighborhood of 350,000,000,000,000,000,000,000 meters per second, and getting faster every second.

This from people who say it's ridiculous to think that we are on a ball spinning at 1000 MPH.

Addressing Flat Earth "Proof"

I will try to address the most common of the flat-Earth arguments here, starting with some I've already mentioned, and saving some of the more bizarre for later.

No matter how high you go, the horizon always looks perfectly flat, 360 degrees around you. Not actually true, but you do have to go very high. The Earth is really, really big compared with anything as small as us, and whatever curve there is, is very subtle indeed.

The horizon always rises to eye level, no matter how high up you go. Just plain false; the horizon is never at eye level, and wouldn't be on a flat plane, either. It may seem so, but that's because the downward angle is so slight, that you cannot tell if you are looking slightly down. The only way to test this is with a transit, and I notice that flat-Earthers don't do this.

Water, we know, always seeks its level. A calm, flat lake, or even the ocean, could never curve around the surface of a globe. If it did, then rivers like the Nile and the Mississippi would be flowing uphill. This is one of those "we all know this is true" statements with no actual proof. Often this argument is accompanied with a horribly out-of-scale drawing showing the effect, often taken directly from Rowbotham, who was absolutely in love with out-of-scale drawings. Water is actually pulled down by gravity, down being toward the center of the planet. Water flows to the lowest elevation relative to the center, which might be up around a curve from a viewpoint outside the Earth, but that's not the same as flowing uphill.

Engineers who build railways and roads never take the curvature of the Earth into account. This is mostly true. But roadbeds aren't perfectly flat. As long as the engineers take care of the local leveling, the gentle curvature of the Earth will take care of itself.

We should not be able to see the Chicago skyline from across Lake Michigan, 90 miles away. This refers to a recent news story from a Chicago station, showing a picture from across the lake, 58 (not 90, or 200 as one flat-Earther claimed) miles away. And it misses the point that this was on the news because it is a rare event, caused by refraction and only under very special atmospheric conditions.

Heliographs, used in the late 19th Century as signaling devices over long distances, should not work on a curved Earth, as the horizon would block the line of sight. A heliograph uses a reflector to send sunlight across long distances and signal an observer many miles away. All kinds of claims are made about heliographs by flat-Earthers, but in fact the longest ever transmission by single heliograph was over a distance of 183 miles, which may seem impossibly long on a curved Earth until you realize that the transmission was made from Mt. Ellen, in Utah (elevation 11,522 feet) and Mt. Uncompahgre, in Colorado (elevation 14,321 feet). High mountain like that can compensate for a lot of curvature.

DuBay talks of a system (consisting of heliostats and limelights) set up along 108 miles of St. George's Channel (the sea channel between Wales and Ireland), but I can find no reference outside of flat-Earth literature, and even within that literature, the implication was that a series of stations was involved, not a single link.

If the Earth were round, then airplanes would have to constantly fly with their noses down to keep from flying off in space. Airplanes, of course, fly perpendicular to the center of the earth, influenced by gravity, and so stay at the same altitude above the globe. Level is not necessarily the same as flat.

The gyroscopic systems used in airplanes would be useless on a globe, since gyroscopes maintain their orientation in space without regard to gravity or curvature. Gyroscopes do maintain their orientation in space, in fact, which is why designers of avionics include gravity-compensated systems in their equipment to account for the curvature. I give you a quote from *Instrument Flying* by David Hoy:

> A gyro, remember, remains rigid in space. A gyro free from any precession placed on your desk would, twelve hours later, appear to be upside-down. In fact, you would be upside-down relative to the gyro and not the other way round. The only way we can use gyros, therefore, is by constant reference to the local vertical by means of a gravity sensor.

If the Earth was spinning at a thousand miles an hour, then cannonballs fired straight up would fall back to Earth miles away from the cannon, instead of nearly perfectly back into the barrel, as happens in reality. First let me point out the the surface of the Earth is traveling over 1000 miles per hour only at the equator. At the poles it travels essentially zero miles per hour. There are two things to remember. The first is that speed doesn't matter; rotation speed does. And the Earth's rotation speed is half that of the hour hand on a clock.

The second thing to remember is that everything on Earth is moving with the Earth wherever it happens to be. So when a plane flies, a cannonball is shot straight up, or a helicopter hovers, it and the very air around it were already moving right along with the Earth. That doesn't change just because you are no longer on the surface.

If the Earth curved away from you, then buildings you see in the distance would visibly lean away from you, and yet no one sees this. Let's go back to the Chicago skyline. Setting aside the fact that, since you are only seeing this mirage because of refraction, the image you are seeing is distorted, let's do a little math. If the skyline is roughly 60 miles away, and the Earth's circumference is roughly 25,000 miles, then that distance accounts for .24 percent of the Earth's complete circle, which is only .864 degrees. Do you really think you could discern if a fuzzy building in the distance is leaning away from you less than one degree?

During some lunar eclipses, the eclipse can be seen starting while the sun is still in the sky. Actually, for every lunar eclipse that ever occurs, somewhere in the world it is possible to see the sun for part of the eclipse. It's in the geometry. It may be a little counterintuitive, but if you actually map it out on a piece of paper, it makes perfect sense. It makes no sense whatsoever, though, on the flat-Earth model. In fact, there is no workable explanation for eclipses, as we see them, on a flat Earth.

When the moon and the sun are in the sky at the same time, the bright side of the moon often does not appear to be in a proper line with the sun. Only if you think the sun and the moon are the same distance away. If the sun and moon were the same size and distance, as the flat-Earth model states, then the juxtaposition of the sun and the moon's lit side would make no sense. But with a fairly distant moon and a very distant sun, the alignment is perfectly explained.

Several constellations can be seen from both the Northern and the Southern hemispheres, which is clearly impossible if the stars are hidden behind the enormous bulging curve of the spherical Earth. The bulging curve of the Earth is not so big when the stars you're looking at are billions or trillions of miles away. Some constellations are, indeed, only visible from one hemisphere or the other, which makes no sense on a flat Earth.

When you see the sun or the moon on the horizon over water, the reflection forms a line from the light source toward you, which should be impossible on a curved surface. This refers to pictures where we see an elongated reflection of the setting sun, which seems to reach from the sun to the shore, aiming right in your direction. Funny thing, though: it only happens when there is a substantial amount of rippling on the surface of the water. Take the same kind of photo on a very still body of water, and the effect goes away completely. It has nothing at all to do with curvature; it's a matter of that rippling water scattering the sunlight in many directions, though the only reflections you can see are the ones that throw light in your direction.

If the Earth is rotating every 24 hours, and orbiting the sun every year, then every six months, day and night should be reversed. This sounds so plausible, but it is astronomically inept. Our days are 24 hours long because that's the amount of time it takes for the Earth to turn to the point where one particular spot facing the sun turns to face the sun again. Because we are orbiting the sun, however, that time is not the amount of time it takes for the Earth to complete one rotation. The way we reckon time creates built-in compensation for the orbit of the Earth around the sun.

Radar works for long distance shots. Radar does use refraction just like light, but like light is has its limitations. This claim usually doesn't come with any data.

Long distance shooters do not take the Coriolis Effect into account. The Coriolis effect is that property of our turning planet that makes free-flying objects moving perpendicular to the axis of the spin appear to swerve to the side (when in fact the ground is turning under them). It makes hurricanes in the Southern Hemisphere tend to turn clockwise and in the Northern Hemisphere turn counterclockwise. The effect is strongest near the poles and non-existent at the equator. Shooters do take it into account if what they are shooting is large enough and goes far enough. Snipers don't really worry about it, as far as my research can tell, but artillery units have books with calculations to compensate for it.

The Coriolis effect affects paper airplanes but not real airplanes I don't know for sure that this is true (paper airplanes are so sensitive to air currents that I don't think any Coriolis effect would be measurable), but if it is, it's because paper airplanes are not flying under power. The Coriolis effect applies to free-flying objects.

Star trails can't work on a sphere. Why not? This is just a bald claim with no facts to back it up. A quick look at the geometry shows that star trails work fine on a globe, and not on a flat plane. Especially when you consider that star trails in the Southern Hemisphere move in the opposite direction from star trails in the Northern Hemisphere.

If the effect of Foucault's pendulum was true, then every pendulum on earth would start by itself and never stop. I don't know where they get this stuff. A Foucault's pendulum does not swing by itself. It changes direction with the rotation of the Earth, depending on your latitude. In fact, it is possible to discern your latitude using only a Foucault's pendulum.

If the Earth was surrounded by a vacuum (instead of,

presumably, by the dome), the vacuum would suck away all the air. We're not talking about a vacuum cleaner, here. A vacuum is just the absence of matter. It doesn't suck away anything.

The globe Earth model requires that the moon lift not only one side of the Earth's oceans, but the entire Earth as well to cause the tide on the opposite side. Yes. So? The Earth and the moon tug at each other all the time. So do the sun and the Earth, and the sun and the moon. Just because someone doesn't understand that doesn't mean that it doesn't happen, or that the Earth is flat.

If the Earth is a ball on an orbital plane, then the Arctic and Antarctic zone should be more alike in temperature, vegetation, and animal life. Except for location relative to the equator, the Arctic and the Antarctic regions have little in common. The Arctic is a frozen ocean surrounded by land, and the Antarctic is ice-covered land surrounded by ocean. As for the animal life and vegetation, climate alone does not determine what kind of plants and animals will inhabit a region; history has a lot to do with it, too.

Dusk and dawn times are different in the Northern Hemisphere than in the Southern Hemisphere. As well they should be, since the Earth is tilted about 24 degrees with respect to the plane of its orbit around the sun, which is what causes the seasons. Perhaps this question is why the dusk and dawn times don't match in cities at the same latitude even if the seasonal difference is taken into account. Which may have something to do with how close a city is to the edge of its time zone. Which may have something to do with the lack of research on the part of flat-Earthers. Because the length of the day in Ensenada, Mexico on December 21st is within a minute of the length of the day in Perth, Australia on June 21st.

All lighthouses are able to defy curve math, as recorded on naval Websites. I see no proof of this, and if anyone has a link with some accurate data, I'd love to see it. I've heard a lot of claims from flat-Earthers about lighthouses being visible from hundreds of miles away, but every mariner I've seen writing on the subject says the furthest any sailor has ever claimed to see a lighthouse from was a bit over 20 miles.

Seeing the *light* from a lighthouse is a different matter; light shining on high clouds above a lighthouse, or light shining through the fog, can be seen from much further away. This is nothing out of the ordinary.

You can watch a ship start to disappear on the horizon and then use a zoom lens or telescope and bring it back into view. If you can still see the ship with a telescope or zoom lens, it hasn't gone over the horizon yet. I've seen YouTube videos of this, and the photographer always shows a wide-angle shot where you can't see the ship, and then zooms in and—amazing!—there's the ship. Which seems to be sitting low in the water. Or maybe the curve is hiding the hull. In any case, the video generally cuts at that point, and we never get to see whether the ship would disappear hull-before-mast with just a little patience.

And I've always wondered how the photographer knew where the ship was well enough to zoom right in on it.

The heliocentric model keeps changing and evolving, and the math and distances change from decade to decade. Yes, that's called "science." It's a self-correcting process, which is why our knowledge of the Universe keeps getting better and better, and why these flat-Earthers have computers and the Internet to spread their nonsense on. Next question.

Sometimes sun rays burst through clouds, showing light hitting us from many angles. These are called "crepuscular rays," and they aren't hitting us from all angles. We're looking straight into them, and perspective makes them appear to diverge. If you take two straight sticks, point them straight out from your head, holding them parallel, and look straight between them, they will seem to diverge. And, under the right lighting conditions, if you look at crepuscular rays and then turn 180 degrees, you can see the rays converge behind you.

The sun seen from high-altitude weather balloons shows a hot-spot on the clouds. If you look at these pictures, you do indeed see a bright spot between the camera and the sun on the tops of the clouds, and this is supposed to show that the sun is not far away, but very close. But if the sun were close enough to case a light on the clouds, there should be two pools of light: one right under the sun, a very large, not very distinct pool, and one like the one we see in the pictures, which is the incidence reflection. But there's only one, because the sun is really, really far away.

You can try this yourself. Go to a hallway with a smooth floor—I have one at home—or you could use a gym or a school hallway. Stand at one end of the hallway and have someone else stand at the other with a lightbulb. Turn out the lights. You will see the pool of light directly under the bulb, lighting your friend's feet. But you will also see the reflection of the bulb itself somewhere between you and your friend. How far depends on the height of your eyes relative to the height of the bulb; if they are near the same height, the reflection should be about halfway between you.

Now, if you're interested, Google angle of incidence. Especially if you're inclined to believe in the flat-Earth.

If the eclipse of the moon is cause by the Earth passing between the sun and the moon, then there should be no full moon; the moon should be in shadow for days every month. Another bit of astronomical ignorance. The moon does not orbit the Earth on the same plane as the Earth orbits the sun. The moon's orbit is tipped about five degrees, which doesn't sound like much, but for an object that is nearly a quarter of a million miles from its shadow source, that's enough to keep it out of the Earth's shadow nearly all the time.

Eratosthenes didn't prove a round earth; as the same answer can be arrived at if the sun is closer and smaller. Eratosthenes didn't actually set out to prove that the world is round; he already knew based on the arguments of Aristotle, among others. Eratosthenes calculated the circumference of the globe, with a fairly small margin of error, somewhere around 300 BC. You *can* conclude that the sun is closer and smaller using the same figures, but then the experiment only works at one set of latitudes. If you were to do the experiment at different latitudes assuming a flat Earth, you would get different sizes and distances for the sun. The fact that the experiment works consistently everywhere indicates a globe.

And, of course, finally, there is the experiment by Rowbotham on the Old Bedford Level, in which Rowbotham clearly saw a boat six miles away which should have been hidden under 16-1/2 feet of curvature. I've already mentioned refraction. Interestingly, a similar experiment was done some years later by a certain scientist who, as chance had it, had been a surveyor in his youth. On a bet (for he was not a wealthy scientist), he set up an experiment with two targets and a telescope. The targets were set at three miles and six miles, thirteen feet above the water to avoid interference from refraction. Three miles from the first target, a telescope was set up on a bridge thirteen feet above the surface of the water and aimed at the far target.

Had the Earth been flat, the targets would have lined up perfectly. But the target at three miles was above the target at six miles, due solely to curvature.

The scientist was the famous naturalist Alfred Russel Wallace, the bet was offered by a flat-Earther named John Hampden, and the whole thing ended up in court, where Russel was judged to have won the wager, though he lost the money because the judge ruled that the wager had been withdrawn.

Flat-Earthers don't talk much about Russel's experiment, though they probably dismiss it because he was, after all, one of the fathers of Natural Selection along with Darwin, and thus untrustworthy in their eyes.

Believe it or not, these are among the more reasonable arguments advanced by flat-Earthers. But the flat-Earthers are, on no account, always this reasonable.

The Lunatic Fringe

Granted, just believing that the Earth is flat puts you pretty far out of the mainstream. But even among the flat-Earthers, there are degrees of madness. This is where things start to get frightening and funny at the same time. Let's take a look at some of the really out-to-lunch things that flat-Earthers say and do.

Some flat-Earthers accuse their detractors of not doing their own research, of just regurgitating what their taught, instead of getting out and doing their own experiments. Ironically, these are often the very people who sit at their computers and do their research by watching other people's flat-Earth videos. There are some who, though sitting at their computers, are doing something more, however. Like flight tracking.

Flight Tracking

I mentioned flight tracking before in relation to GPS, but for other reasons, airline flights in the Southern Hemisphere are a problem for the flat-Earth model. A direct flight from Sydney, Australia, to Santiago Chile takes about 12-1/2 hours direct. But some calculations on the AEP, if taken as if it represented a real flat Earth, would extend the 7,500-mile distance between those two cities (by the shortest path on a globe) to nearly 12,500 miles (by shortest path on a flat plane). So there's another reason that flat-Earthers track those flights: to show that they are faked.

So you'll see videos making a big deal about flights dropping off the online flight trackers when flying between southern continents, no matter how many times flat-Earthers are told that this is just to be expected, and that it happens in the Northern Hemisphere, too. But the videos keep coming.

Speaking of flights, the flat-Earthers all got very excited when a news story broke about a woman giving birth on an airplane en route to Los Angeles. I think that's pretty exciting, too, but for the same reason most of us would think so. But the flat-Earthers got excited because they think it proves that the Earth is flat. The doctor who delivered the baby got on the flight in Bali. The woman who had the baby got on in Taipei. About a third the way through the flight, the woman's water broke. The pilot diverted to Fairbanks, Alaska, but the baby had already been delivered by the time they got there. And, the flat-Earthers ask: why Alaska? If you look on the map, Hawaii is right on the way. But, on the "flat Earth map" (the AEP), Alaska is on the way. Aha!

Well, not really. If you look on a Mercator projection, then yes, Hawaii seems to be right on the way. But the shortest route from Taipei To L.A. is past China, Russia, and, yes, Alaska, on a globe. The fact that the majority of this journey took place in the northern hemisphere means that the AEP gives a pretty good approximation.

But this incident keeps showing up as proof that the Earth is flat, and furthermore that all the airline pilots know it. Funny that all those airline passengers who saw coastline out of the windows on the left side didn't think too much about it. They must be brainwashed.

One YouTuber was obsessed with the Samoan islands. He looked all over for a direct flight between Apia, in Samoa, and Pago Pago, in American Samoa, which are only 75 miles apart. Well, he didn't actually look all over. He looked at aggregator sites that sold only tickets from the major airlines, and so found flight that went way out of the way, to cities in the Northern Hemisphere, like Los Angeles. He saw this as evidence that Samoa and American Samoa are actually much further away from each other than we've been told.

Commenters told him, over and over, that there were no two major airlines with contracts for both airports, and that direct flights were available through a small island-hopper airline (easily found online), but he just called us all trolls and didn't seem at all interested in looking for himself. He was just too happy to have found his little smoking gun.

Another thing I can't quite understand about flat-Earthers is how resistant they are to facts. Time and again, flat-Earthers will show flight paths that "don't make sense on a globe," and then use a flat Mercator projection to show the flight path, ignoring anything close to what the actual path is over a sphere.

In Search Of the Edge

For awhile, there were a slew of postings of parts of the Canadian Film Board documentary *In Search Of the Edge,* which begins with the tale of one Andrea Barnes, a woman obsessed with finding the edge of the flat Earth, who travelled to Antarctica, where she disappeared. The film then proceeds to lay out the case for the flat Earth, with on-camera narration by scientists and professionally-produced animated sequences.

After someone pointed out that *In Search Of the Edge* is fiction, purposely made to look like a documentary to order to teach critical thinking skills to Canadian students, the references to it died down a bit (though many said that they knew it all along, but that the filmmakers were slyly trying to slip in the flat-Earth "truth" on the pretext of making an educational film). You can still find people posting this on YouTube, however, as if it were completely true.

Flat Earth Maps Everywhere

Another YouTuber got excited (they do that a lot) when he found a map in the collection of the Boston Public Library. The map is an azimuthal equidistant projection with a North Pole origin, and it bears the title: "Gleason's New Standard Map Of the World." And under the title are the words "As It Is." Another smoking gun, no doubt. Not only do the powers that be know that the world is flat, but they issued Gleason patents in several countries.

Seems pretty cut-and-dried, doesn't it? Until you investigate further. First, looking at the map, you see that is also contains the words. "Based On the Projection Of J. S. Christopher, Modern College, Blackheath, England." So, if it's a projection, it's not really a flat-Earth map, is it? At the upper left, you can find another little blurb that says, "Time and Longitude Calculator." Hmm. Maybe that's what that little piece of something attached to the middle of the map is for. Seems to be torn off. This is, after all, a scan of the original, and the original is from 1892. There must be a story behind this map.

Indeed, there is. You can download your own copy of the map and find the patent for it with a quick Google search. You'll discover that the little something in the center was a chart, one of two (one is missing) used to help school children calculate time zones, which was actually a relatively new concept at the time. No one ever claimed that this was a true map of the flat Earth. Until now, anyway.

Flat-Earthers, especially those deeper into the conspiracy aspects of this, make much of the fact that the AEP is used, in part, in the design for the official flag of the United Nations. Missing the ice ring, for some nefarious reason. It can't be, I suppose, that it just makes a nice graphic design and, unlike a depiction of a globe, can show all the nations in a single, flat image while still making a nice circle (graphic designers love circles; they fit everywhere).

No, that's too easy.

Flat-Earthers love to find examples of the flat Earth "hiding in plain sight." They latch onto every sighting of a flat-Earth reference in the movies, like Jack Sparrow opening something that looks like the AEP in *Pirates Of the Caribbean*. In the opening shot of *Back To the Future* (following the Universal logo, which of course has that globe for indoctrination), one of Doc Brown's clocks is based on the AEP (as are many 24-hour clocks). The logo for the Disney series *Ducktales* has a world graphic loosely based on the AEP. Disney is supposed to be knee-deep in this flat-Earth conspiracy, because, well, who knows?

The appearance of the AEP in *Cast a Giant Shadow, Twelve Monkeys, Mars Attacks,* and possibly (but not definitely) in *Minghags,* is seen as a clue planted to hint at the reality of the flat Earth. Even the hologram that the controllers use to monitor the arena in *Hunger Games* and the running gag of the Pizza Planet truck appearing in Pixar movies is seen as evidence that Hollywood is in on the conspiracy and is sending subliminal messages.

Globes Everywhere

Of course, the flip side of seeing AEP maps everywhere is seeing globes everywhere, and looking upon it as a means of indoctrination. I won't even bother to give examples. You know that globes are everywhere; they are among our most basic tools for seeing our world as it is.

Subtle Flat-Earth Clues

Some flat-Earthers think they see evidence everywhere they look that governments and even entertainers know about the flat-Earth and are subliminally trying to tell us. One YouTuber made an entire video about a photoshoot that David Bowie did in 1975 for his album *Station To Station* where he's lying on the floor drawing the tree of life. In one image on the backdrop there is a scrawl of circles where the video's poster swears he sees a flat plane with the heavenly dome in the middle of the tree of life. Go find the image and tell me what you see. Bowie, unfortunately, having died of cancer in 2016, is unavailable for comment.

Another YouTuber did an extensive analysis of the lyrics

and cover art of Pink Floyd, and found numerous flat-Earth references which seem to have escaped everyone else, including a claim that "The Wall" is a reference to the giant Antarctic ice wall.

There was one video that claims that Hollywood knows about the flat Earth just because a radio announcer at the beginning of the movie *Porky's* used the phrase "the stars are secure in the firmament"!

Another YouTuber posts episodes from James Burke's *Connections* series and claims that they, too have flat-Earth clues in them. The posts are taken down because of copyright infringement, leading the YouTuber to claim that these documentary episodes have been "banned," and giving the aura of some conspiracy to hide them from us, even though you can buy them as a boxed set on Amazon.com.

The Matrix is very popular with some flat-Earthers, not just as entertainment, but because they believe we actually live inside of The Matrix. It's all an illusion. How they know that outside of this illusions lies a flat plane is never actually explained.

And then there's *The Truman Show*. If we're not living in The Matrix, maybe we're all living in a glass dome controlled by a director and his crew somewhere, and not allowed to venture beyond the dome, which is protected by that nasty Antarctic ice wall. There are people who actually believe this.

Some of the people with thoughts along these lines also think that the sun, moon, stars, and planets are holograms projected on the dome. How they expect this to work geometrically (or even in any way related to how holograms work) is never explained.

Other flat-Earthers try to find hidden meanings though numerology, going a bit nuts whenever any variation of 666 shows up, or the "Masonic" number 33. These numbers are obviously planted as clues to help us find the "truth" of the flat Earth.

All Those Freemason Scientists

As I mentioned, the conspiracy to hide the flat Earth and promote the globe is ascribed to various secret and not-so-secret grounds, but no one seems to get more attention than the Masons.

Pretty much any scientist who has ever contributed to our understanding of the globe is labeled as a Freemason. I won't say accused, even though that is the spirit in which the flat-Earthers offer it. In some cases the label is correct. In others it's far off the mark. But the Masonic connection to the flat-Earth conspiracy is a big part of the litany. This is from Eric DuBay:

> **From Pythagoras to Copernicus, Galileo and Newton, to modern astronauts like Aldrin, Armstrong and**

> Collins, to director of NASA and Grand Commander of the 33rd degree C. Fred Kleinknecht, the founding fathers of the spinning ball mythos have all been Freemasons! The fact that so many members of this, the largest and oldest secret society in existence have all been co-conspirators bringing about this literal "planetary revolution" is beyond the possibility of coincidence and provides proof of organized collusion in creating and maintaining this multi-generational deception.

Beyond coincidence. And more than a little beyond belief since Pythagoras died about 1800 years before Freemasonry got started. This is proof #191. If this is the quality of proof that is typical of flat-Earth "theory," then it's no wonder the globe model is dominant.

So intense can be the obsession with Freemasonry that flat-Earthers will post videos of people making perfectly ordinary gestures, and interpret them as Masonic hand signs. Commenters such as myself are labeled as Freemasons just because we challenge the flat-Earth evidence with our annoying facts.

Aside from calling every scientist who supports the globe

(virtually every scientist, that is) a Mason, flat-Earthers seem to hold particularly intense hatred for certain scientists. Copernicus and Galileo. Newton and Kepler. Einstein and Hawking. Sagan and Tyson.

Neil DeGrasse Tyson gets called out especially, since he is such a current public figure in astrophysics. He gave an interview recently where he was talking about the scale of surface features of Earth compared to its size, and during the interview he mentioned, a bit jokingly, that the Earth is wider around the equator then from pole-to-pole, and that this bulge is a bit south of the equator, making it a little pear-shaped. The flat-Earthers jumped right on that, creating pictures of the globe map in the shape of a pear, citing it as evidence that the pictures of the Earth are faked because they look perfectly round, and making endless fun of Dr. Tyson, a man who's accomplishments and mental capacity they can't ever dream of.

And posting the interview on YouTube with its ending cut off, where Dr. Tyson said that the real lie about the Earth is that it has any surface features at all; from a cosmic standpoint, it is a perfect sphere.

Flat-Earthers get a kick out of tearing down famous scientists. They're fools, or dolts, or tools of the machine, or liars, Masons, or shills. Never mind that they have made the world as we know it possible.

Flat Earth Experiments

Not all flat-Earthers just sit around at their computers looking at other people's stuff. Some of them do experiments. Sort of. I haven't seen a recreation of the experiment at the Old Bedford Level yet (much less Wallace's version), but there are people who get out there with hand-held cameras, point them at objects of indeterminate height and distance, see all (or usually only part) of those objects, and declare the Earth flat. Controls and accurate data or math are nowhere to be found.

More hilariously, one YouTuber held up an apple, poured water on it, and declared that since the water fell off, the oceans could not stick to a spherical Earth. What's really funny is that for reasons having nothing to do with gravity, there was a film of water on the apple which, in scale, is deeper than the deepest oceans on Earth. Somehow, flat-Earthers think this proves something.

Many flat-Earthers have decided that one way to prove the flatness of the Earth is to show that the moon shines by its own light. Part of this is the idea that moonlight is actually cold (DuBay even says this in proof #132, along with other dubious old wives' tales about the moon). So they set up thermometers under the full moon, one in the moonlight and one shaded from the moon, and lo! and behold, the one in the shade is a little warmer. How could that be?

Let's consider how much light the moon casts toward the

Earth. The moon only reflects 11% of the light that it receives, and it's much, much smaller than the sun. So, it doesn't send us much light, certainly not enough to get a rise out of a thermometer. Now, during the day, the sun is warming up the ground. When darkness falls, the ground starts releasing some of that warmth back to the air. In the open, the warm air is gong to rise away from the ground. Under a shade, the warm air will get trapped under whatever you're using to shade the thermometer. Thus the temperature difference in and out of the shade. Nothing to do with moonlight at all.

One flat-Earther, much to his credit, did the same experiment under the new moon and got the same result as under the full moon. He remains convinced of the flat Earth, but no longer claims that moonlight is cold.

Another well-known flat-Earther set up a laser on one beach across a bay from a camera at another beach. The distance was a little over four miles. According to this young man, the laser should have been invisible under 10 feet of curvature. And yet, the camera could see the laser from across the bay. Slam-dunk, right?

Hardly. First, the math was, again, based on the eight-inches formula I've mentioned before, and that keeps getting repeated as if it were correct. It doesn't take observer height into account. The height of the observer was given as 30 inches, but that was actually the height of the tripod. The tripod was well back from the water on the beach, making the camera perhaps as much as two feet higher. The height of the laser was given as six inches, but again, it wasn't in the water, it was on the beach. Add to that the fact that, close to the water, as Rowbotham knew, refraction is a big factor.

And the laser's specs for power, beam diameter, and divergence were never given. The fact that the laser light could be seen from four miles shouldn't be a surprise to anyone.

The one that raises my hackles, probably more than any other, is not so much an experiment as an attempt at a demonstration. I've already mentioned how flat-Earthers insist that sunsets are a matter of perspective, that the sun doesn't actually sink below the horizon, but merely shrinks away. This is their attempt to prove this.

I have seen this more than once: a YouTuber will set up a table, with a camera right at the surface level (actually a little below). They will hold a coin on the top of the table near the camera, and then they will slide it back and forth *on the surface of the table* to show how much it looks like the sun setting below the horizon. So, what they're saying is that the sun is *on the ground?* And that you and I, when witnessing sunsets, are standing in a hole with our eyes just below the surface of the Earth?

Nothing you can say to these people will convince them

what a horrible analogy this is.

Because it's in my area of expertise as a photographer and filmmaker, I will devote an entire chapter to perspective just a little later in the book. But I mention this little demonstration here as a further example of flat-Earthers' idea of what constitutes a viable experiment or demonstration.

Nikon Coolpix p900

The Coolpix P900 is an amazing camera. It's also kind of weird to someone who's been doing photography from the time when even automatic exposure was an expensive luxury. It has an extremely long zoom ratio of 83 times. And so flat-Earthers have been using it to zoom in on distant skylines and ships at sea to "prove" that there is no curvature. I expect that, just as I expect them not to understand what they're looking at.

But some flat-Earthers have been using it as if it were an astronomical telescope, even though it's missing one important feature that all astronomical telescopes have: manual focus. And so, when you point it at a light in the sky and zoom in, it has no frame of reference for focusing that point of light, and produces an image that can only be described as an indistinct blob. And, since the image is taken through moving air, the point of light is likely to be rendered as a color-changing, undulating, indistinct blob.

That's okay; I doubt the folks at Nikon designed the

camera for stargazing. The problem is that some flat-Earthers taking video of these color-changing, undulating, indistinct blobs, are claiming that that's what the planets and stars actually look like. The camera does not lie, right? Further, this is offered as evidence that the stars and planets are not what they seem. What they are instead isn't fully explained, though one YouTuber claims they are angels, making no connection to the flat Earth, save for some reference to the dome.

Fake Spaceflight

I could write an entire book addressing the false claims made by moon hoaxers, but aside from the fact that others have already done a better job of that than I ever could, I truly don't think I could stomach the research involved.

But you don't have to go back to the 60s to find NASA footage for the flat-Earthers to attack. In January 2016, astronaut Scott Kelly, aboard the International Space Station, did a wonderful demonstration with two hydrophobic paddles and a drop of water, which he used to play ping-pong between the paddles. And, of course, the anti-NASA crowd came out of the woodwork (or out from under their rocks) to proclaim it a fake. Hollywood green screen (they love to claim that green-screen can do anything, even through few have any idea how it works), and computer-generated imagery (CGI, apparently, is all-powerful).

One commenter, who claims to have been doing CGI since 1998, says that it's obviously CGI because it Kelly's eyes fail to exactly follow the water drop as it moves back and forth between the paddles. What's obvious to me, with 45 years of filmmaking experience, is that, with nobody operating the camera, Kelly keeps checking his framing in the monitor. His eyes move frequently offscreen, just as happens with most people when they are taking video selfies, and then back to the water droplet, which he looks directly at, when he needs to keep the droplet under control.

The other thing my filmmaking experience tells me is just how impossible the claims of fakery are. Seriously, back in the days of the moon landing, faking the broadcast from the moon would have been a far more amazing feat than the moon missions themselves. And now, the International Space Station transmits a 24-hour-a-day live feed, which shows the goings on inside the station when the crew is on duty, with all the effects of microgravity there for all to see, and pictures of the Earth from space, where verifiable weather patterns can be seen, when the crew is off-duty.

That's an impossible amount of CGI. The same can be said of Japan's Himawari-8 geostationary satellite, which sends a new picture of the Earth from it's location facing Papua New Guinea every ten minutes. Now, flat-Earthers (and other anti-NASA conspiracy theorists) like to point out that these, like the images from DISCOVR, the US National

Oceanic and Atmospheric Administration satellite that orbits the sun, are admitted by the space agencies to be composites. But that doesn't mean they are fake. It's important to remember that satellites are sent into space to do a job, not take pretty pictures. Himiwari-8 and DISCOVR take images through different filters for scientific analysis of weather and climate patterns. By combining the red, green, and blue images into one, a full-color image is created.

It's also important to remember that your own digital camera does the same thing; the only difference is that it takes the three images simultaneously, whereas the satellite cameras take them in series (which makes sharper individual images).

But the "evidence" of NASA fakery gets even weirder. Many have speculated that micro-gravity footage from space is created in airplanes, called reduced-gravity aircraft, which climb and dive in precisely-calculated paths to create different reductions in gravity. The problem with this idea is that these flights only create about 25 seconds of microgravity for every 65 seconds of flight, and the crew of the space shuttle is on camera, in long, continuous shots, for far longer than that.

That has some flat-Earthers speculating that NASA has created some kind of anti-gravity system to simulate spaceflight here on Earth. Which is actually kind of an odd thought for a group of people who pretty much need to discount gravity altogether to support their world view. One YouTuber put up footage of sky diving simulators, which are essentially vertical 200MPH wind tunnels, and claimed that these are proof of NASA having the means to fake microgravity.

I guess all of those live feeds from the ISS are dubbed to eliminate the incredible wind noise from one of these things.

Antartica

As I've mentioned, Antartica is a big problem for the flat-Earthers. If it is supposed to be a giant ring around the disc, then anyone traveling there would be able to figure that out. So the flat-Earth gospel includes a prohibition on travel to Antarctica, as laid out in the Antarctic Treaty. United Nations troops, supposedly, patrol the perimeter and will shoot anyone trying to enter unescorted.

The first joke here is anyone patrolling the perimeter of what the flat-Earthers say is Antartica, an ice ring of 75,000 miles in circumference. It's another one of those physical impossibilities that's just never explained.

About 37,000 people visit Antartica each year. Granted, they are escorted, for reasons of both safety of environmental protection, but they flock there. What do the flat-Earthers say to that? Well, more fakery, they say. The tourists never actually venture to the pole, they just come in on one section of the ice ring, get shown some ceremonial pole, and are fooled. I guess they are also fooled by the Antarctic midnight sun, which is impossible on a flat plane. But I digress.

Flat-Earthers leverage the harsh conditions and the Antarctic Treaty to claim that the land is inaccessible and carefully controlled. They like to tell the story of Jarle Andhøy, who went on an illegal voyage to Antartica and was arrested for it. But they don't like to tell the whole story; indeed, few of them have bothered to research it at all. Here's a short report on the incident from *The Old Salt Blog*:

> Last February, Jarle Andhøy and Samuel Massie set off from the yacht Berserk II in McMurdo Sound on the Antarctic coast in an attempt to drive ATVs towards the South Pole. Shortly thereafter, in an Antarctic storm, the EPIRB [emergency position indicating radiobeacon] from the Berserk II was activated. Following an extensive search by the New Zealand Coast Guard and private ships, the only sign of the yacht or the three crew aboard was debris and an empty life raft. Andhøy's ill-fated expedition was mounted without authorization or

insurance. He later paid a fine of 25,000 NOK to the Norwegian Polar Institute.

So, Andhøy wasn't actually arrested as such, and he certainly wasn't stopped at the ice wall; he and Massie were driving all-terrain vehicles toward the pole, without sufficient support. They probably would not have been caught had the yacht not gotten into trouble and sent a distress signal. As for punishment, the fine didn't even come close to covering the cost of rescue operations. 25,000 krones is less than 3,000 US dollars.

Of course, flat-Earthers make much of failed expeditions to cross Antarctica, failing to mention (or learn) that these are attempts to be the first to do something like cross on skis, or do a solo cross. Logistical companies like Antarctic Logistics & Expeditions, out of Salt Lake City, Utah, would, I think, be very surprised at the notion that one can't cross from one coast of Antartica to another, since they do so routinely.

Unless, of course, they are in on the conspiracy.

Perspective For Flat-Earthers

To me, one of the dead giveaways that we do not live on a flat plane is the fact that the sun and the moon rise and set beyond the horizon. In every model I've yet seen of a flat Earth, the motion of the sun and moon cannot ever match what we see in real life. As I mentioned, the flat-Earth explanation for this is "perspective."

I'm a filmmaker and photographer, and I've been paid to direct professional animation artists during my career. This is kind of up my alley. So, let me give you a quick overview of what perspective really is, and why it cannot cause the illusion of the sun setting and rising over a flat Earth.

First off, perspective is not a scientific concept. Flat-

Earthers will talk about the "law of perspective," but never give a source. Perspective is a set of tools used by artists to attempt to recreate the experience of a three-dimensional world using two-dimensional media, like drawings and paintings. Photographers and filmmakers have less control than other artists over perspective, but we still have to deal with it (since a photograph is still two-dimensional). And we learn to use it compositionally, and manipulate it in various ways, to achieve certain effects.

I'm not going to get into that much detail here, though, because it has nothing to do with sunsets and the flat-Earth.

There are two kinds of perspective: linear and aerial. Aerial perspective has to do with the way atmospheric distortion affects how we see things in the distance, and how using different colors and levels of detail can create this effect in a work of art. That's not what the flat-Earthers are talking about. They're talking about linear perspective, which you learned about in grade school by drawing things like railroad tracks and square buildings that reached for a distant vanishing point.

But in real life, there's only one guiding principle related to perspective: things look smaller when they are further away from you. That's it. That's the whole thing. Flat-Earthers will tell you that the ground "rises up" to meet the horizon, and that the sky "comes down." But what's really happening is that, as things are further from you, they occupy a smaller percentage of your field of view. The ground seems to "rise up" because you can't see through it. The sky does not "come down."

Everything gets smaller, and everything gets smaller proportionally, in all directions, at the same rate. The distance between objects and the objects themselves follow the same progression. So let's use the moon rising and setting as our example, because the moon is not bright enough to cause substantial glare, and you can view it safely without the need of special filters.

If you watch the moon during its traverse across the sky from moonrise to moonset, it might appear to be a little bigger when it's close to the horizon than when it's high in the sky. There are actually a couple of theories as to why this is true, but it doesn't actually matter at this point. For most of the time you see the moon, it's roughly the same size, that is, it occupies the same amount of space in the sky.

It doesn't get smaller as it sets, and even if you see it over the ocean, where there are no trees or hills in the way, you will see it slowly sink beneath the horizon. No matter how high up you go, on the tallest building or the highest mountain, it will always appear to sink down beneath the horizon.

Now, let's think about the flat model based on the AEP. Let's pretend that the flat-Earth model is real. Let's pretend that you are on the beach near Coaque, Ecuador, very close to the equator. The beach faces west at that location. It's midnight on March 21st, and the full moon is right over your head. According to flat-Earth astronomy, it's about 3000 miles above you, and circling at about 1600 miles per hour (if the AEP were true, rather than a distortion, the equator would be over 37,000 miles in circumference).

To ensure that you're facing west, you pull out the compass, find north, adjust for the difference between true and magnetic north, and turn 90 degrees to your left. What's going to happen to the moon? It will move past you going roughly west. But because it's traveling in a circle over your head, it won't stay to the west; it will veer to the north. It will appear to get smaller, gradually, and it will appear to slow down.

Six hours after midnight, when you expect to see the moon sinking into the ocean directly due west, the flat-Earth model shows it precisely 45 degrees to our right, exactly northwest. Not only that, but it's now 9000 miles away from you, three times as far as it was before, moving over the skies just north of the Samoan islands. It should appear to be one third the size it did at midnight. And it will still be 19.5 degrees above the ocean.

It will never set. Six hours later, it will be directly north of you. It will be over Sumatra, and over 12,000 miles away. It will still be 14 degrees above the ground (at that point), and still big enough to see.

Then? It will start coming back, growing larger, getting closer, moving faster, until it's midnight again.

And everything I just said just covers perspective. It doesn't take phases, eclipses, or what portions you might be seeing of the surface of the moon during its traverse into account.

This is how I know that the flat-Earth model is nonsense. The sun, the moon, rising, setting. It's simple.

Flat-Earthers tell me that I don't know anything about perspective, that I don't understand human perception, that some magic version of perspective is at play that's different from the kind I've been using for fifty years since I first picked up a camera.

But not one has come close to providing anything that resembled evidence.

Flat-Earth Morals

Flat-Earthers claim, almost universally, to be purveyors of some great truth, to be piercing the veil of lies forced onto us by governments and secret societies. As such, they are trying to claim the high moral ground. But their behavior often reveals a different picture.

As I've mentioned, flat-Earthers will often resort to the most vile attacks on anyone who dares to enter their domain (even their public YouTube videos) and hold a different view on the shape of the Earth. But if heresy is a trigger for hostility, think what apostasy brings. One YouTuber who was trying to come up with a flat-Earth map that reconciled flight times with the distances on the map.

After much wrangling, which even had him creating one map with two Australias, he finally gave up, and decided that the world is probably not flat, after all. The backlash was horrifying. He has been shunned, his name now associated with the worst kinds of shills and trolls. He is no longer welcome in any way in the community he was an active member of. All for the sin of doing what flat-Earthers keep insisting we should all do: doing his research, and trusting his observations.

Flat-Earthers don't seem to care whose feelings they hurt, or whose memories they sully. You may have seen memes around the Internet where people used facial recognition software to find very old pictures that are dead ringers for modern celebrities. Well, a NASA hoaxer used the same technique to find living people whose faces resembled those of the astronauts killed in the Challenger and Columbia disasters, and has used those pictures to claim that those two tragic events were faked, and that the astronauts are alive and well. I don't know if the originator of this disgusting display was a flat-Earther, but the flat-Earthers have definitely picked it up and used it in their videos many times.

A beloved president from the past, John Kennedy, was not spared the indignity of the flat-Earth zeal either. Multiple videos have claimed that Kennedy was assassinated because he was about the reveal the secret of the flat Earth.

One of them quoted this portion of a speech that Kennedy delivered to the American Newspaper Publishers Association in April of 1961. Note that this was 2-1/2 years before his assassination, when the President had only held office for a few months.

> I want to talk about our common responsibilities in the face of a common danger. The events of recent weeks may have helped to illuminate that challenge for some; but the dimensions of its threat have loomed large on the horizon for many years. Whatever our hopes may be for the future–for reducing this threat or living with it–there is no escaping either the gravity or the totality of its challenge to our survival and to our security–a challenge that confronts us in unaccustomed ways in every sphere of human activity.
>
> This deadly challenge imposes upon our society two requirements of direct concern both to the press and to the President–two requirements that may seem almost contradictory in tone, but which must be reconciled and fulfilled

if we are to meet this national peril. I refer, first, to the need for a far greater public information; and, second, to the need for far greater official secrecy.

The very word "secrecy" is repugnant in a free and open society; and we are as a people inherently and historically opposed to secret societies, to secret oaths and to secret proceedings. We decided long ago that the dangers of excessive and unwarranted concealment of pertinent facts far outweighed the dangers which are cited to justify it. Even today, there is little value in opposing the threat of a closed society by imitating its arbitrary restrictions. Even today, there is little value in insuring the survival of our nation if our traditions do not survive with it. And there is very grave danger that an announced need for increased security will be seized upon by those anxious to expand its meaning to the very limits of official censorship and concealment. That I do not intend to permit to the extent that it is in my control.

Very powerful words, even removed from the context of the speech. But flat-Earthers have found hidden meaning in word like "illuminate" (implying something to do with the Illuminati), and "dimensions" and "horizon" and "sphere," to say nothing of the phrase "there is no escaping the gravity." And the fact that this speech mentions secret societies, to the author of this tripe, suggests that Kennedy was killed by NASA and the Masons.

Very recently, another man's dignity was under attack by certain members of the flat-Earth community. On January 24th, 2016, explorer and former British Army officer Henry Worsley, died of complications from peritonitis, from a bacterial infection he contracted while attempted a solo crossing of the Antarctic continent. He was following in the footsteps of explorer Sir Ernest Shackleton. Worsley got further than Shackleton, calling for an airlift only 30 miles short of his goal. For his effort, he raised more than £100,000 for wounded British soldiers.

But some flat-Earthers on YouTube have taken his story as proof that the Antarctic continent does not exist, and even suggested that the Crown killed Worsley to keep him from revealing the secret.

And, just as I was getting ready to put this book to bed, one of only twelve men to have ever walked on the moon, Edgar Mitchell, died. I should have learned it from the news, or from one of the scientists whom I follow on Twitter. But as the luck of timing would have it, I learned of it from this tweet:

> Any person sad over the death of #EdgarMitchell must feel empathy for lying frauds. He NEVER WENT TO THE MOON. #MoonLandingHoax #FlatEarth

This kind of behavior takes these flat-Earthers out of the realm of serious debaters, and puts them in the camp of fanatics with no moral compunction.

Why It Matters, Revisited

The cover story of *National Geographic Magazine* for March 2015 is entitled "The War On Science." There has always been resistance to some of the ideas that science produces, especially when it runs head-on into religious dogma. But given the vast benefits of scientific inquiry, science-bashing has usually been tempered by reason and critical thinking.

But something has changed. The Internet, that boon of communication that has given us so much, has also made it easier for outliers to find each other, to reinforce each other's beliefs, and to give many the anonymity that emboldens them to express uncomfortable, unpopular, or even outrageous ideas.

And while it's good to be open to new avenues of investigation, in an age of overwhelming information, it's also important to practice discretion, to filter out the nonsense to find the real evidence. I see no sign of that filtering process in the flat-Earth world. Every little perceived chink in the globe model's armor is seen as a complete victory, with no further investigation, even if the chink is actually nothing but a complete misunderstanding, or misrepresentation.

For not all of the flat-Earthers are in it for the knowledge. Some seem to have made themselves a nice chunk of money from promoting this viewpoint, and others, while not making any income, are getting feelings of power and influence that they would never have achieved in another realm. This is not a great motivation for hardheaded evaluation of evidence, and certainly not for admitting one's mistakes.

Another aspect of the flat Earth is that it's heavily entangled with the "truther" community. Not that all truthers believe that the Earth is flat; in fact, as I mentioned, some think the flat-Earth movement itself is some kind of government operation designed to discredit truthers by associating them with the crazy flat-Earth notions. But I've seen very few flat-Earthers who aren't also strong proponents of some other conspiracy theory (besides the conspiracy to cover up the flat Earth, and the moon landing hoax, of course). This can make it as hard to disengage from the flat-Earth community when doubt creeps in as it is for most people to leave their churches.

This is assuming, of course, that the flat-Earther in question really believes what they are promoting. Another consequence of the ubiquitousness and anonymity of the Internet is what is known as Poe's Law. To quote Wikipedia:

> **Poe's law is an Internet adage which states that, without a clear indicator of the author's intent, parodies of extreme views will be mistaken by some readers or viewers for sincere expressions of the parodied views.**

In other words, someone who's actually trying to make fun of the flat Earth can come across as someone who believes in the flat Earth, unless he or she says something specific to show that it's a joke.

Then there are trolls, people who post inflammatory comments and videos just to get a reaction (the expression comes from a method of fishing, trolling, that involves dragging bait through the water; the connection to the mythical Scandinavian creatures came later, but it does seem to fit).

And, further, there are sock puppet accounts, accounts on social media that are used to make a subject appear more popular than it is. Many flat-Earthers use multiple accounts for both posting and commenting, and there are also what I call "search result spam" accounts, who take videos on a subject when it becomes popular, steal them and re-upload them to make money.

And, who knows, there might actually be shills on either side of the flat-Earth exchange (I won't dignify it by calling it a debate), people who are paid to promote a particular point of view whether they believe it or not. But I kind of doubt it.

The point of all this is that it's really hard to tell just how many people genuinely think that the Earth is flat. It would be good to know just what the extent of the damage is. But I think perhaps that matters less than the proliferation of material being posted on the Internet, sincere or not, because that kind of exposure is bound to rope in a number of people who are innocent, ignorant, gullible, or just angry enough at the powers-that-be to buy into it.

Especially when a celebrity (albeit a minor one) supports the idea publicly. Recently, the rapper B.o.B. and reality TV star Tila Tequila posted on Twitter that they accept the flat-Earth. Their influence is limited, and they took a lot of heat for their views, but fans of the two celebrities might consider the idea if it comes from someone they like more readily than if it just showed up on the Internet, say in the comments on a video or photo gallery on the Apollo mission (which does happen—flat-Earthers are tenacious).

So, we get back to why it matters. Cinematographer and filmmaker S.G. Collins did a wonderful video about why, in 1969, it was impossible to fake the moon landings, and at the end he talks about why it matters. I suggest you watch his excellent video, but let me paraphrase: it's about what Collins called "the fate of knowing," seeing the difference between what you can know and what you wish for. The urge to believe, he said, drives people to exchange a piece of their soul for the comfort of being a rebel. That applies perfectly to flat-Earthers.

Collins goes on to say that clinging to beliefs against all the evidence of your rational mind blinds you to the real conspiracies that are being perpetrated against you every day. He mentions the Iraq war, the financial bailout, the Patriot Act, and the right to indefinite military detention without charge. These things are real. The flat Earth is not.

I don't believe that the flat-Earth is a government psychological operation any more than I believe that the globe is a Masonic plot to separate the masses from God and hide vast resources from the public. But if someone did come up with a distraction to get us to ignore the various ways they are really messing with us, this one would have been pretty good.

But it's a distraction. There's more important work to be

done, including understanding our world through the practice and process of science, and helping people understand science and technology for real.

That's not a job I'm well-qualified for; I'm not a working scientist, I'm just a writer, filmmaker, and songwriter.

But I can do my part, in my own little way, by doing my best to head off the nonsense. I hope you'll join me in that endeavor.

About the Author

Gordon Brooks is a writer, a filmmaker, and a songwriter. Once upon a time, he was one of the pioneers of interactive media, working for Philips Media as a technical director, sound editor, writer, producer, and director. He now lives in New Hampshire, USA with his wife and three sons. He can be reached at me@gsbrooks.com, and you can see more of his work at gsbrooks.com.

See the latest on the flat Earth at Gordon's new blog, The Earth Is Not Flat! at embracetheball.blogspot.com.

Notes

The AEP map illustration is by Daniel R. Strebbe, and is from August 15, 2011. It is licensed under the Creative Commons Attribution-Share Alike 3.0 Unported license. It was not created as a flat-Earth map, but merely an azimuthal projection.

The illustration of the Bedford Level Experiment is from *Zetetic Astronomy* by Parallax, which is in the public domain.

"Moon Hoax Not" by S.G. Collins can be found on YouTube for your viewing pleasure.

Zetetic Astronomy and *100 Proofs the Earth Is Not a Globe* are both in the public domain, and can be found on many sites.

Eric DuBay's ebook *200 Proofs the Earth Is Not a Spinning Ball* is available for free download from his site, AtlanteanConspiracy.com

Made in United States
North Haven, CT
30 January 2025

65168633R00064